REV. DR. COPP'S DISCOURSE

ON THE

ATLANTIC TELEGRAPH.

THE ATLANTIC TELEGRAPH: AS ILLUSTRATING THE PROVIDENCE AND BENEVOLENT DESIGNS OF GOD.

A

DISCOURSE,

PREACHED IN THE

BROADWAY CHURCH, CHELSEA,

AUGUST 8, 1858,

BY

JOSEPH A. COPP, D. D.

BOSTON:
PRESS OF T. R. MARVIN & SON, 42 CONGRESS STREET.
1858.

The Atlantic Telegraph Cable was announced from Trinity Bay, Thursday, August 5th, and on the next Sabbath morning, the 8th, was the subject of the following Discourse to a joint congregation of the Broadway and Plymouth Societies, in the Church of the former; and at the earnest request of members of both, is now given to the public.

DISCOURSE.

JOB XII. 9.

THE HAND OF THE LORD HATH WROUGHT THIS.

It is said that the first telegram sent over the experimental wire of Morse, and which was expressive alike of the good taste and piety of the inventor, was the appropriate Scripture, "What hath God wrought?" A text of similar import has been selected this morning, to discourse, in the way of religious improvement, on the last and grandest achievement of the electro-telegraph.

The subject which, at the present time, justly absorbs public attention throughout Great Britain and the United States, with demonstrations of universal joy, is the union, so happily established between the two countries, by the electric cable. Incredulity may no longer doubt—fears and misgivings are at an end—the marvelous work is done! The wonders of art and practical science in the past, are as children's work—the creations of the nursery—when compared with this.

A few years ago, the electric telegraph was a thing unknown, a thing incredible to general science; but now it has become a matter of common experience, and of every day business. But to-day we witness another step in the wonderful art, more wonderful than any of its former triumphs—the crowning miracle of all. A few days ago, what was characterized as visionary and impracticable, under natural hinderances deemed insurmountable, and what was pronounced by some eminent scientific men impossible on scientific grounds, is to-day a certain, a pleasing fact. In the face of all reasonings and fears to the contrary, behold the reality! That bold work, against which mighty nature seemed to hurl defiance—to proclaim her prohibition in tempests and stormy waves—is done! Angry ocean may foam and rage, and skeptical science may hesitate and doubt, but the cable *is laid!*

An epoch worthy of commemoration and thankfulness dawns on us this holy day—worthy of most religious recognition, as coming from God. To-day, mother England and her American daughter, heretofore separated for nearly three hundred years of time, and nearly three thousand miles of distance, are brought to shake hands across an annihilated ocean. A proximity of communion and intimacy is henceforth established between them, like that between the sister municipalities, in the midst of which it is our privilege to live. It is true, we cannot to-day hear the bells of Old England ring out her church-going people, but we might hear, along the wonderful cable, her Christians pray and her ministers preach, and lift their pious acclamation over this new and heaven-sent bond of union between the Christian millions of the two countries. Wonderful event! As we contemplate it, the heartfelt utterance rises spontaneously to our lips, "THE LORD HATH WROUGHT THIS!" An event so wonderful in itself, and so prophetic of good to mankind, must be ascribed to the wonder-working providence of Him "who sitteth on the circle of the earth"—who "laid the foundations thereof," and "stretched the line upon it" of a far-reaching benevolence.

THE PROVIDENCE OF GOD IN THE ACCOMPLISHMENT OF THE OCEAN TELEGRAPH, AND THE GOOD TO MANKIND IT MUST WORK, is the subject of our special consideration.

The Christian cannot doubt that this interesting event, from first to last, is of God. It may indeed be the fruit of genius, courage and unwearied toil. We would gladly ascribe all honor to the inventor, an illustrious son of Massachusetts, whose work confers glory on his country and his native State. But who inspired the fortunate train of thinking, which has led to the grand result? Who put it into the heart of the immortal Morse, to begin and maintain those tedious, expensive, and often disappointed experiments, through years of neglected toil and discouraged hope, which have to-day culminated in the last—greatest wonder of the world? The scientific discoverer himself, was the first to give the glory to God. Like Bezaleel, the son of Uri, he was "filled with the Spirit of God, in wisdom and understanding, and in knowledge." His soul was touched by a spark brighter than electric light — a spark of heavenly wisdom itself — and the truth burst forth upon him. In that first dispatch, which his own hand sent over the speaking wire, he demonstrated the reality of the wonderful invention, and laid its honor, at the same time, at the footstool of God.

But let us pass from a truth, which atheism alone will deny, and contemplate that particular providence under which the Ocean Telegraph was finally accomplished.

Twelve months ago, when those two mighty ships, with their escort, went forth to do a work by which continents were to be joined, universal expectation was high and sanguine. Science and skill, it was supposed, had so nearly conquered every opposing difficulty, that the enterprise could hardly fail. Great, therefore, was the public disappointment, in the disaster that terminated the expedition. A year rolls round, and all the appliances of science and toil, within the reach of human power, have done their best to correct every error and imperfection, and to perfect the arrangements for a second and successful trial.

But God, to whom belongs, and should be ascribed, the glory of the work, will teach men their weakness and dependence; will make them feel, that in an enterprise on such a magnificent scale, and of such important moral bearings, success "is not of him that willeth, nor of him that runneth, but of God that showeth mercy." For a great work, men must be disciplined. Nothing great is accomplished in this world by human agency, as all history will show, without a previous training. So God will teach the leading minds in this great enterprise, and an interested public looking on, that success depends on His favor, who holds the winds in his fist, and the waters in the hollow of his hand. He had planted a mighty ocean between the continents, and before this stormy barrier of six thousand years shall be forced by modern art, and yield to the embracing nations, His permission must be obtained, and the world be made to know, that He consents to the union, and smiles on the work.

Previous failure and disappointment, had somewhat cooled the ardor of expectation, and schooled the minds of men; but the work of discipline was not complete. And now comes the final trial. Preparëd for every contingency, as was supposed, the ships turn their prows to mid-ocean. But the winds of heaven are let loose against them, and the angry billows threaten with destruction the daring fleet. God, who measures out the tempest for discipline, and not for destruction, restrains its violence. The ships outlive the storm. One but just survives it. With thanksgivings to God for sparing mercy, they meet at the appointed place, and, with no little despondency, begin the work assigned them. We know the unhappy issue of another, and another trial. Deeply discouraged they return to Ireland, and by the failure of the undertaking spread discouragement over the civilized world. But the enterprise for which so much of mind, and labor, and money had been expended, cannot be abandoned. At least another effort must be made, to satisfy the public, and to sustain a noble company, and if possible a sinking stock, which had become almost worthless in the market.

Under these clouds of discouragement, the cable fleet set about making their last desperate trial. It again set sail, like a forlorn hope, braving the dangers of the fatal breach. The ships, henceforth ever memorable, return to mid-ocean. They return, as the world now felt, to the closing disaster —to throw away their precious cable, seal the financial ruin of the company, and stamp the whole enterprise as a visionary, daring, and impracticable thing. All men looked with sympathy on the noble band of scientific adventurers, and naval officers, doomed to be the victims of the failure.

But another scene awaits us, in which we find an illustration of the Christian proverb, that "man's extremity is God's opportunity." The time, ordained in the eternal counsels of wisdom, had come for success. The managing minds, the men designing and executing, were now disciplined. Science and toil had done all that was possible, and were humbled. And now it was deeply felt, and frankly acknowledged, if God be favorable, it may be done—without a special providence it must fail. And here let us turn aside, and take a look into a secret chapter of this memorable expedition. What were the feelings and encouragements of those practical men, on whom rested the executive responsibilities? Of those on board the Niagara, we can only speak with certainty. These men were rebuked and humbled by past failure, and had come to commit the great work to God in prayer. Their own complete arrangements, were less an encouragement than dependence on the blessing of God. The commander of the ship, and the chief electrician, felt so profoundly the need of God's blessing, that they sought it through the prayers of God's people on shore, and humbly implored it in their own state-rooms on board. They invested their undertaking with a high moral importance, rising far above scientific achievement and commercial convenience, as looking to a future of love and good will, of peace and religious benefit to mankind; and thus, purified from low aims, they were prepared to commend the undertaking to Him whose purpose it is, that all things shall so work, as finally to make purity and virtue triumphant.

It is most interesting to know, that while the world of commerce and ambition,—while bankers and philosophers, cabinets of state, and faculties of science, were contemplating the affair from their own stand-points, and in the light of their own peculiar interests and pursuits, there were others, on the land, and in the cabins of the Niagara, a chastened band, praying over it, in the serene light of philanthropy and religion.

Heaven is not unconcerned, when men of pure intentions and humble confidence, call for an exercise of grace, or interposition of Providence in behalf of a worthy object. Under the lofty impulses of religion and humanity, God has given to the soul a large liberty of asking. "Where two or three (said the Saviour) are gathered together in my name, there am I in the midst of them;" and if these "shall agree, as touching anything they shall ask, it shall be done for them of my Father which is in heaven."

Such were the circumstances of moral discipline and of dependence, under which the two ships meet and make their last splice. No human power was equal to the contingencies of the undertaking. Disappointment had prepared them for the worst; and now, having committed all to God's holy providence, these two ships separate, and hopefully, prayerfully, but tremblingly, bid each other farewell.

And now a great and joyful surprise is about to break upon the world. "Weeping may endure for a night, but joy cometh in the morning." Had it been announced the other day, as the final dispatch, that the last effort had been unsuccessful, and the cable was lost, a feeling of regret, indeed, would have followed the announcement, yet no one would have been disappointed; it was already lost in the fears of the multitude. But the time had come for the stupendous work to receive the approval of God, that man might see the divine hand in the accomplishment, and his glory in the end.

And now the surprise that has gladdened the country, and thrilled the hearts of millions in America and Great Britain, is one, under the circumstances, that points emphatically to heaven, and declares, in the language of the Psalmist, "This is the Lord's doing, and it is marvelous in our eyes."

And now let us proceed to inquire, as the conclusion of the whole matter, WHAT IS THE GRACIOUS END CONTEMPLATED BY PROVIDENCE IN THIS WONDERFUL EVENT, AND WHAT GOOD TO MANKIND WILL IT WORK?

An event in which the hand of God has been visible, must of course be for good. All things, however, are providential, in that they are wisely employed or overruled, and, directly or indirectly, move forward towards a great and benevolent end. The world, and all things in it and concerning it, are subject to an almighty will, and shall work out the purpose of infinite benevolence. God's ultimate object cannot be defeated; it will be accomplished as certainly as there is omniscience to plan, and omnipotence to execute.

But in the view of man, some things are more obviously and directly providential than others. While God is actually in all events, yet in many of them he moves invisibly and mysteriously; but in others, his agency unveils itself to a conscious and awe-struck world. Thus Job said, "I have heard of thee with the hearing of the ear, but now, mine eye seeth thee." Is it not so in the event before us? When all hope of success, under obstacles apparently insurmountable, had been well nigh abandoned, God reveals his own arm; he holds back the winds; he calms the troubled sea; he lays, as it were, with his own hand, the slender wire over that dark and mysterious plateau, hidden for unknown ages, in the depths of ocean, "that it might take hold of the ends of the earth."

Now what is the design of a providence so signally displayed? We may say generally, and with confidence, some good of corresponding greatness and mercy to the children of men.

First, this telegraphic cable is to be a pledge and bond of peace. Two great Christian nations of the same origin, of the same language, and embarked on the same enterprises of civilization and humanity, ought never to contend but in the noble rivalry of doing good. To contend in low, brutal, wicked war, let us believe, will henceforth be impossible to these kindred nations. They have fought their last battle, and shed the last drop of fraternal blood. The bond of an everlasting union and sympathy is laid between them; it is the cord of love, "a threefold cord, not quickly broken." It will bind the heart of Old England with noble, fraternal beatings, to that of Young America. Along this cord of connection, will flow those moral forces which, like the processes of life, excite to combine, and, conveying a healthful influence to every part, will serve to unite in harmony and strength, the general system. Through this wonderful medium, will pass and repass the utterances of commerce, of letters, of friendship and religion. Every pulsation of business, of science, and of society along the electric chain between the millions of either land, will strengthen the amity of the countries.

Acquaintance, is the practical philosophy of forbearance and love. Experience shows, the more we understand ourselves and others, the more improbable becomes hurtful difference and vulgar contention. Nations, like individuals, need but mutual acquaintance, to discover the common advantages which flow from peace and friendship. Wars among nations, like quarrels among individuals, are blunders and mistakes. Justice and humanity, no more than true policy, forbid them in every case; and it never can be the interest of men or of nations, to disregard their claims.

Let us therefore hail the success of the great experiment, as the pledge of increasing intimacy and acquaintance, and consequently of peace and good neighborhood between nations destined to control the empire of the world. In the struggle of the past, it was the distance of the Colonies, that lost them to the mother country. Their wrongs were unknown, their cries unheard, their just demands unappreciated. The mother, afar off, became indifferent, and then cruel; and the daughter rebellious, and then independent. The result, in its present form, is very well; but henceforth all contention is needless, as it would be wicked between the nations. We therefore hail the success of the Ocean Telegraph, as closing forever the gates of Janus between kindred millions, and pledging them to a long career of peace and prosperity. Let them go on in an alliance of friendship and love, to do the work of humanity and civilization for the less favored members of the family of mankind.

But, secondly, in this event, we see an omen of promise for Christian progress. Merchants and capitalists projected the Atlantic Telegraph, with worldly views and for worldly ends. Selfishness, very likely, labored to accumulate the wealth, that first moved in the mighty enterprise; and selfishness, it may be, has employed the means and agencies that have accomplished it. But God knows how to overrule the worldly views and ends of man, and turn the plans and labors of the selfish to the account of his kingdom. Commerce builds ships for her own gains and glory, but Religion employs them to send the Bible and the Missionary to the heathen. Capitalists lay railroads, but along the iron track, Christianity and its literature, and its thousand appliances for good, travel and are diffused. The improvements of art, and the facilities of modern traffic, are the ready avenues through which the grace of God pours the riches of mercy on a sinful world. And so in the case of this sub-Atlantic Telegraph; for whatever advantage of state, of commerce, or of letters, it may be regarded and used, God will employ it for the higher purposes of religion and humanity; along it will burn many a dispatch of mercy, to quicken the pulse of Christian activity and holy love of souls, in the old or new world. Transatlantic piety and American zeal, kindled up by intercommunicate flashes, drawn from the battery of heaven itself, will be felt reciprocally and simultaneously in both hemispheres. Wonderful vehicle! Like the spinal cord in the animal system, which conveys the impressions and volitions, from the seat of intellect in the brain, to the most distant members, deciding with instantaneous precision the most remote and complicated movements; so this international spinal connection will serve to convey and multiply the impressions of the divine will, to parts at hand, and to places afar off. Over it will run the living word of Christ, with the speed of lightning, the warmth of love, and the certainty of truth, to perform its saving work on either side of the globe. It is the fulfillment of prophecy—in a significant sense, "It is the voice of him that crieth in the wilderness (of waters), Prepare ye the way of the Lord, make straight in the desert (of deep ocean) a highway for our God."

Thus, science and wealth have unwittingly laid their contributions at the feet of our blessed Redeemer, to be employed in the promulgation of that saving truth, for which he lived, suffered, and died, and now liveth forever more. This grand achievement belongs to Christ,—to the glory of his kingdom and the progress of his truth we would consecrate it this day. He, to whom God has given all things, rightly claims the telegraph as his own. It is the angel of the Apocalypse, having the everlasting gospel to preach; the angel of heaven's own lightning, hurrying with the speed of thought,—not through the clouds above, but, under depths, mysterious, profound, and awful, carrying the message of God, "Peace on earth and good will to men," with angelic certainty and dispatch to the nations.

Thus we may rejoice to-day as Christians and philanthropists, in an event which will bring glory to God and good to man every where, and which will contribute in its future operations to that promised and long prayed for epoch, when Jesus shall reign universally, and

"His kingdom stretch from shore to shore,
Till moons shall wax and wane no more."

www.ingramcontent.com/pod-product-compliance
Lightning Source LLC
LaVergne TN
LVHW020638110826
845149LV00004B/1256

9781418190835